INDUSTRY STANDARD
OF THE PEOPLE'S REPUBLIC OF CHINA

Standard for Rock and Soil Classification of Railway Engineering

TB 10077-2019

Prepared by: China Railway First Survey and Design Institute Group Co., Ltd.
Approved by: National Railway Administration
Effective date: August 1, 2019

China Railway Publishing House Co., Ltd.

Beijing 2022

图书在版编目(CIP)数据

铁路工程岩土分类标准:TB 10077-2019:英文/中华人民共和国国家铁路局组织编译. —北京:中国铁道出版社有限公司,2022.11
ISBN 978-7-113-29369-7

Ⅰ.①铁… Ⅱ.①中… Ⅲ.①铁路工程-岩土分类-行业标准-中国-英文 Ⅳ.①TU4-65

中国版本图书馆 CIP 数据核字(2022)第 110119 号

Chinese version first published in the People's Republic of China in 2019
English version first published in the People's Republic of China in 2022
by China Railway Publishing House Co., Ltd.
No. 8, You'anmen West Street, Xicheng District
Beijing, 100054
www. tdpress. com

Printed in China by BEIJING CONGREAT PRINTING CO.,LTD.

© 2019 by National Railway Administration of the People's Republic of China

All rights reserved. No part of this publication may be reproduced or transmitted in any form or by any means, electronic or mechanical, including photocopying, recording, or by any information storage and retrieval systems, without the prior written consent of the publisher.

This book is sold subject to the condition that it shall not, by way of trade or otherwise, be lent, resold, hired out or otherwise circulated without the publisher's prior consent in any form of binding or cover other than that in which it is published and without a similar condition including this condition being imposed on the subsequent purchaser.

ISBN 978-7-113-29369-7

About the English Version

To promote the exchange and cooperation in railway technology between China and the rest of the world, National Railway Administration organized the translation of this Standard.

This Standard is the official English language version of TB 10077-2019. The Chinese version of this Standard was issued by National Railway Administration and came into effect on August 1, 2019. In case of discrepancies between the two versions, the Chinese version shall prevail. National Railway Administration owns the copyright of this English version.

China Railway Economic and Planning Research Institute Co., Ltd. and Beijing Times Grand Languages International Translation and Interpretation Co., Ltd. prepared the English version. Beijing Jiaotong University provided great support during review of this English version.

Your comments are invited and should be addressed to Technology and Legislation Department of National Railway Administration, 6, Fuxing Road, Beijing, 100860, P. R. China and China Railway Economic and Planning Research Institute Co., Ltd., 29B, Beifengwo Road, Haidian District, Beijing, 100038, P. R. China.

Email: jishubiaozhunsuo@126.com

The translation was performed by Cao Ce, Zhang Lixin, Dong Xuewu, Wang Ke and Liang Chao.

The translation was reviewed by Chen Xi, Wang Lei, Li Nufang, Fei Jianbo, Yuan Chunyan and Ju Xiaoqiang.

Notice of National Railway Administration on the Issue of the English Version of Six Railway Engineering and Construction Standards including *Code for Design of Suburban Railway*

Document GTKF [2021] No. 46

The English version of six railway engineering and construction standards including *Code for Design of Suburban Railway* (TB 10624-2020) is hereby issued. In case of discrepancies between the Chinese version and the English version, the former shall prevail.

China Railway Publishing House Co., Ltd. is authorized to publish the English version of these standards.

List of the English Version of Six Railway Engineering and Construction Standards

S/N	Title	Reference No.
1	Code for Design of Suburban Railway	TB 10624-2020
2	Code for Design of Railway Industrial Siding (Provisional)	TB 10638-2019
3	Standard for Rock and Soil Classification of Railway Engineering	TB 10077-2019
4	Specification for Engineering Test of Train Dispatching and Commanding System and Centralized Traffic Control System	TB/T 10435-2020
5	Specification for Engineering Test of Railway Computer Based Interlocking System	TB/T 10436-2021
6	Specification for Engineering Test of Chinese Train Control System	TB/T 10437-2021

National Railway Administration

December 17, 2021

Notice of National Railway Administration on the Issue of Railway Industry Standards
(Engineering and Construction Standard Batch No. 1, 2019)

Document GTKF [2019] No. 19

Eight railway industry standards (see Table 1) including *Code for Engineering Geology Investigation of Railway* (TB 10012-2019) are hereby issued and will come into effect on August 1, 2019. Seven railway industry standards (see Table 2) including *Code for Geology Investigation of Railway Engineering* (TB 10012-2007) are withdrawn.

China Railway Publishing House is authorized to publish these standards.

Table 1   List of Standards Issued

S/N	Title	Reference No.
1	*Code for Engineering Geology Investigation of Railway*	TB 10012-2019
2	*Standard for Rock and Soil Classification of Railway Engineering*	TB 10077-2019
3	*Code for Railway Tunnel with Gas*	TB 10120-2019
4	*Technical Specification for Test of Foundation Piles in Railway*	TB 10218-2019
5	*Technical Specification for Safety Control of Blasting Operation in Railway Engineering*	TB 10313-2019
6	*Code for Supervision of Railway Construction Project*	TB 10402-2019
7	*Inspection Specification for Structure Concrete Strength of Railway Engineering*	TB 10426-2019
8	*Code for Dynamic Acceptance of Mixed Traffic Railway*	TB 10461-2019

Table 2   List of Standards Withdrawn

S/N	Title	Reference No.
1	*Code for Geology Investigation of Railway Engineering*	TB 10012-2007
2	*Code for Rock and Soil Classification of Railway Engineering*	TB 10077-2001
3	*Technical Code for Railway Tunnel with Gas*	TB 10120-2002
4	*Technical Specification for Testing of Railway Piles*	TB 10218-2008
5	*Code for Construction Project Management of Railway*	TB 10402-2007
6	*Inspection Specification for Structure Concrete Strength of Railway Engineering*	TB 10426-2004
7	*Guidance on Dynamic Inspection of Completion Acceptance for Mixed Passenger and Freight Railways Construction*	Document TJS [2008] No. 133

National Railway Administration
April 18, 2019

Foreword

Standard for Rock and Soil Classification of Railway Engineering unifies the criteria for rock and soil classification of railway engineering and plays the supporting role as a basic standard. This Standard is prepared based on revisions made to *Code for Rock and Soil Classification of Railway Engineering* (TB 10077-2001), geotechnical experience of China railway in recent years and relevant Chinese and foreign research achievements.

This Standard consists of 4 chapters, namely General Provisions, Terms and Symbols, Classification of Rock and Rock Mass, and Classification of Soil.

The main revisions are as follows:

1. The basic principles for classification of rock, rock mass and soil are specified.

2. The contents on salinastone, saline rock and basic quality classification of rock mass are added.

3. The classification of rock formation thickness is adjusted.

4. The particle size grouping of soil and classification of gravel are revised.

5. The grading of clay with medium compressibility is refined.

6. The quantitative judgment indexes of compactness of gravel are added.

7. Criteria are added for the vane test and flat dilatometer test of soft soil.

8. The classification of sensitivity of soft soil and the judgment indexes of static cone penetration test for the soft soil are revised.

9. The criteria of saline soil are revised.

10. The seasonally frozen soil and the frost-heaving grading of seasonally thawed soil layer are added.

We would be grateful if anyone finding the inaccuracy or ambiguity while using this Standard would inform us and address the comments to China Railway First Survey and Design Institute Group Co., Ltd. (No. 2, Xiying Road, Xi'an, Shaanxi Province, 710043), and Planning and Standard Research Institute of National Railway Administration (Block B, Jiangong Building, No. 1, Guanglian Road, Xicheng District, Beijing, 100055) for the reference of future revisions.

The Technology and Legislation Department of National Railway Administration is responsible for the interpretation of this Standard.

Prepared mainly by:

    China Railway First Survey and Design Institute Group Co., Ltd.

And also by:

    China Railway Liuyuan Group Co., Ltd.

    Southwest Jiaotong University

    China Academy of Railway Sciences Corporation Ltd.

Drafted by:

    Ju Xiaoqiang, Liu Xiaodong, Meng Xianglian, Li Xiang, Shu Lei, Du Shihui, Wang Chunlei, Wu Xiyong, and Zhang Qianli.

Reviewed by:
Gu Xiangsheng, Xu Zailiang, Xiao Daotan, Du Wenshan, Wang Yongguo, Meng Qingwen, Cao Huaping, Yu Peng, Zhou Shiguang, Yang Changsuo, Yang Pengjian, Liu Haozheng, Liu Chun, Liu Xun, Dai Ying, Li Hongjiang, and Chai Guanhua.

Record of updates:
Code for Rock and Soil Classification of Railway Engineering (TB 10077-2001).

Contents

1 General Provisions ·· 1
2 Terms and Symbols ·· 2
   2.1 Terms ·· 2
   2.2 Symbols ·· 2
3 Classification of Rock and Rock Mass ·· 4
   3.1 Classification of Rock ··· 4
   3.2 Classification of Rock Mass ··· 6
4 Classification of Soil ·· 13
   4.1 General Requirements ·· 13
   4.2 Classification of General Soil ·· 13
   4.3 Classification of Special Soil ··· 17
Words Used for Different Degrees of Strictness ··· 28

1 General Provisions

1.0.1 This Standard is prepared with a view to unifying the technical criteria for rock and soil classification of railway engineering.

1.0.2 This Standard is applicable to the rock and soil classification of railway engineering.

1.0.3 The rock and soil classification shall be conducted by combining field identification with laboratory and field tests, and by combining qualitative classification with quantitative assessment.

1.0.4 Rocks shall be classified according to their hardness, softening property and weathering resistance. The rock shall be designated as special rock when it contains special mineral composition or structure, has special physical, mechanical and chemical properties, and affects the engineering geological conditions.

1.0.5 The rock mass shall be classified or zoned according to the structural type, rock formation thickness, joint development degree, degree of influence of geological structure, integrity, weathering degree and basic quality of rock mass.

1.0.6 Soil shall be classified according to the accumulation age, geological genesis, shape and gradation of soil particles or plasticity index. The soil shall be designated as special soil when it contains special mineral composition or structure, has special physical, mechanical and chemical properties, and affects the engineering geological conditions.

1.0.7 The classification of rock and soil not involved in this Standard may comply with the relevant Chinese standards in force.

2 Terms and Symbols

2.1 Terms

2.1.1 Rock

A single rock block not containing distinct structural planes.

2.1.2 Rock mass

The natural geological body of rock with considerable volume within the affected area of the project.

2.1.3 Structural plane

The geological interface of planes, seams, layers and bands with certain direction, size, shape and characteristics in the rock mass.

2.1.4 Basic quality of rock mass

The inherent basic property of rock mass affecting the rock mass stability that is determined by the rock hardness and rock mass integrity.

2.1.5 Special soil

The soil containing special mineral composition or structure, having special physical, mechanical and chemical properties, and affecting the engineering properties of soil.

2.1.6 General soil

The soil having no special structure or properties, containing no special composition or containing special composition that does not affect the engineering properties of soil.

2.2 Symbols

$a_{0.1-0.2}$ — coefficient of compressibility;
BQ — basic quality index of rock mass;
$CEC(NH_4^+)$ — cation exchange capacity;
C_u — undrained shear strength;
D_r — relative density;
e — natural void ratio;
F_s — free swelling ratio;
I_L — liquidity index;
I_P — plasticity index;
I_r — liquidity-plasticity ratio;
I'_r — liquidity-plasticity limit ratio;
I_D — soil type index;
k — permeability coefficient;
K_f — weathering coefficient;
K_p — wave velocity ratio;
K_r — softening coefficient;
K_v — index of rock mass integrity;

M—montmorillonite content;

N—measured blow counts of standard penetration test;

$N_{63.5}$—measured blow counts of heavy dynamic penetration test;

N_{120}—measured blow counts of extra-heavy dynamic penetration test;

P_s—specific penetration resistance of static cone penetration test;

P_p—swelling force;

q_c—cone tip resistance of static cone penetration test;

R_c—saturated uniaxial compressive strength of rock;

S_r—degree of saturation;

S_t—sensitivity;

S_u—vane shear strength;

S'_u—remoulded vane shear strength;

q_u—unconfined compressive strength of undisturbed soil;

q'_u—unconfined compressive strength of remoulded soil whose density and moisture content are the same as those of undisturbed soil but whose structure is completely destroyed;

v_p—longitudinal wave velocity;

w—natural moisture content;

w_A—total moisture content;

w_{sa}—water saturated absorptivity;

w_L—moisture content at liquid limit of soil, referred to as liquid limit;

w_p—moisture content at plastic limit of soil, referred to as plastic limit;

W_u—organic matter content;

α_w—water content ratio;

δ_S—collapsibility coefficient;

$\overline{DT}$—average salt content of soil;

δ_0—average thaw subsidence coefficient;

η—average frost-heaving ratio.

3 Classification of Rock and Rock Mass

3.1 Classification of Rock

3.1.1 The qualitative classification of rock hardness may be made according to Table 3.1.1.

Table 3.1.1 Qualitative Classification of Rock Hardness

Name		Qualitative identification	Representative rock
Hard rock	Very hard rock	With crisp sound, rebounding and hand-shocking sense in hammering; hard to break and with no water absorption reaction after immersion	Unweathered or slightly-weathered granite, basalt, gneiss, diorite, quartzite, siliceous limestone, siliceous cemented sandstone or conglomerate, etc.
	Hard rock	With relatively crisp sound, slight rebounding and slight hand-shocking sense in hammering; relatively hard to break and with slight water absorption reaction after immersion	Moderately-weathered very hard rock; unweathered or slightly-weathered ignimbrite, marble, killas, dolomite, limestone, calcite-cemented sandstone, etc.
Soft rock	Moderately soft rock	With no crisp sound and rebounding in hammering; relatively easy to break and with obvious water absorption reaction; with scratches with fingernail after immersion	Highly-weathered very hard rock; moderately-weathered hard rock, unweathered or slightly-weathered phyllite, mica schist, sandy mudstone, calcite and argillaceous-cemented siltstone and conglomerate, marlstone, shale, tuff, etc.
	Soft rock	With deep sound and no rebounding in hammering; with dents, easy to break and can be split with hands after immersion	Highly-weathered very hard rock; moderately-weathered-highly-weathered hard rock; moderately-weathered soft rock and unweathered or slightly-weathered argillaceous rocks; mudstone, argillaceous-cemented sandstone and conglomerate, etc.
	Very soft rock	With deep sound and no rebounding in hammering; with deep dents and can be split with hands; can be kneaded into lumps or crushed after immersion	All kinds of completely-weathered rocks; highly-weathered soft rock; rocks with poor diagenesis; Tertiary sandstone, mudstone and other hypabyssal rocks

3.1.2 The quantitative classification of rock hardness shall be made by adopting the saturated uniaxial compressive strength of rock (R_c), and shall comply with Table 3.1.2.

Table 3.1.2 Corresponding Relation between R_c and Rock Hardness Classified Qualitatively

Saturated uniaxial compressive strength of rock R_c (MPa)	$R_c > 60$	$60 \geqslant R_c > 30$	$30 \geqslant R_c > 15$	$15 \geqslant R_c > 5$	$R_c \leqslant 5$
Hardness	Very hard rock	Hard rock	Moderately soft rock	Soft rock	Very soft rock

3.1.3 The classification of softening property of rock by the softening coefficient shall comply with Table 3.1.3.

Table 3.1.3 Classification of Softening Property of Rock

Name	Not easily softened	Easily softened
Softening coefficient K_r	>0.75	≤0.75

Note: The softening coefficient K_r is the ratio of the saturated uniaxial compressive strength to the uniaxial compressive strength of the same rock under dried state.

3.1.4 The weathering resistance of fresh rock may be classified according to Table 3.1.4.

Table 3.1.4 Classification of Weathering Resistance of Fresh Rock

Item	Classification		
	Not easily weathered		Easily weathered
Softening property	Not easily softened		Easily softened
Frost resistibility	Frost resisting		Not frost resisting
Magmatite structure	Fine-grained		Coarse-grained
Rock-forming minerals	Mostly of quartz	Mainly of feldspar, pyroxene and hornblende	Mainly of pyrite, olivine and biotite
Cementitious substance	Siliceous	Calcareous	Argillaceous
Weathering resistance time	Not obviously weathered after exposure for one or two years		Weathered after exposure for several days to months

3.1.5 The rock containing many hydrophilic minerals that experiences a large volume change when the moisture content changes shall be judged as swelling rock.

The judgment of swelling rock shall be made according to the requirements of investigation stage, the field geological characteristics and the laboratory test indexes of swelling rock, and as per Table 3.1.5-1 and Table 3.1.5-2.

Table 3.1.5-1 Field Geological Characteristics of Swelling Rock

Topographic features	Low undulating hills with a relative height of 20 m - 30 m are formed generally. The hill top is rounded and the slope surface is round and smooth, with a gradient of less than 40°. There are wide U-shaped valleys between the hills. Scarps are usually formed when sandstone interlayers are present
Lithology	There are mainly grayish white, grayish green, grayish yellow, purplish red and gray mudstone, argillaceous siltstone, shale, weathered marlite, weathered basite magmatic rock, montmorillonite tuff and rocks containing anhydrite and mirabilite. The rocks are made up of fine particles with slippery feeling when exposed to water. The argillaceous swelling rock is mainly distributed in Carboniferous, Permian, Triassic, Jurassic, Cretaceous or Tertiary strata
Structure	The rock strata are mostly in thin and medium-thick layered form, and fissures are developed. The fissures are mostly filled with grayish white or grayish green substances rich in montmorillonite
Weathering	The weathered fissures develop further along the structural plane and bedding plane, making the rock blocks cut by the structural plane become more broken. The rock fragments on the surface are weathered into chicken manure soil that is peeling off obviously. In natural water-bearing rocks, micro-fissures often occur along the bedding direction when exposed to sunlight. The dry rock is easily disintegrated into fragments, debris or soil after being soaked in water

Table 3.1.5-2 Laboratory Test Indexes for Swelling Rock

Test item		Index
Not-easily-disintegrated rock	Free swelling ratio $V_H(\%)$ or $V_D(\%)$	$V_H(\%)$ or $V_D(\%) \geq 3$
Easily-disintegrated rock	Free swelling ratio $F_S(\%)$	$F_S \geq 30$
Swelling force P_p (kPa)		$P_p \geq 100$
Water saturated absorptivity $w_{sa}(\%)$		$w_{sa} \geq 10$

Notes: 1 The not-easily-disintegrated rock shall be judged by the larger one of the axial free swelling ratio V_H or radial free swelling ratio V_D;

2 The easily-disintegrated rock shall be crushed and sieved through a sieve of 0.5 mm to remove coarse particles, and then tested by adopting the test method of free swelling ratio of soil;

3 When two or more items meet the indexes listed in this table, the rock may be judged as swelling rock.

3.1.6 The rock with the following characteristics shall be judged as salinastone:

1 Pure salt crust or salt deposit formed in Quaternary period.

2 Mudstone or sandstone rich in soluble or moderately soluble salt minerals in the sedimentary rock strata formed before Quaternary period.

3 Evaporative chemical sedimentary rock with layered distribution of gypsum, anhydrite, mirabilite, halite, tronite trona, etc.

3.1.7 The salinastone may be classified into chloride salinastone, sulfate salinastone and carbonate salinastone according to the main chemical composition and as per Table 3.1.7.

Table 3.1.7 Type of Salinastone

Type of salinastone	Representative salinastone categories
Chloride salinastone	Halite ($NaCl$), leopoldite (KCl), calcium salt ($CaCl_2$), magnesium salt ($MgCl_2$), etc.
Sulfate salinastone	Gypsum ($CaSO_4 \cdot 2H_2O$), anhydrite ($CaSO_4$), natural mirabilite ($Na_2SO_4 \cdot 10H_2O$), glauberite ($Na_2SO_4 \cdot CaSO_4$), etc.
Carbonate salinastone	All kinds of tronite trona (Na_2CO_3, $NaHCO_3$), monohydrate sodium carbonate ($Na_2CO_3 \cdot H_2O$), heptahydrate sodium carbonate ($Na_2CO_3 \cdot 7H_2O$), decahydrate sodium carbonate ($Na_2CO_3 \cdot 10H_2O$), etc.

3.1.8 For the sedimentary rock strata covered or intercalated with salinastone, when the content of soluble salt distributed in the rock mass in the forms of film layer, concretion, fiber and screen mesh, as well as the earthy soluble salt filled in fissures exceeds 0.3% due to chemical infection or leaching, the sedimentary rock shall be judged as saline rock.

3.2 Classification of Rock Mass

3.2.1 The classification of rock mass by the structural type shall comply with Table 3.2.1.

Table 3.2.1 Classification of Rock Mass by Structural Type

Name	Geologic body type	Shape of main structural body	Development of structural plane	Engineering characteristics of rock mass	Potential geotechnical problems
Integral massive structure	Homogeneous massive magmatic rock or orthometamorphite, hugely-thick layered sedimentary rock or para-metamorphic rock	Massive block or very thick layer	Mainly of primary structural joints and layers, which are mostly of closed type, with a spacing of structural planes being greater than 1 m, with no more than 1-2 groups of structural planes, and without falling rocks caused by dangerous structural planes	With high overall strength, the rock mass is stable and it may be regarded as a homogeneous and elastic isotropic body	Partial sliding or collapse of unstable structural body, or rock burst of deep underground cavity

Table 3.2.1(continued)

Name	Geologic body type	Shape of main structural body	Development of structural plane	Engineering characteristics of rock mass	Potential geotechnical problems
Blocky structure	Blocky magmatic rock or orthometamorphite, thick layered sedimentary rock or para-metamorphic rock	Thick layered, blocky or columnar	Only a small number of joints and fissures with good penetrability are found. The spacing between structural planes is mostly greater than 0.4 m, and there are generally 2-3 groups of structural planes, with a small number of detached bodies	The overall strength is high, the structural planes are mutually restrained, the rock mass is basically stable and close to the elastic isotropic body	Partial sliding or collapse of unstable structural body, or rock burst of deep underground cavity
Stratified structure, block stone and crushed stone-like structure	Multi-rhythm thin and medium-thick layered sedimentary rock or para-metamorphic rock	Blocky stone, crushed stone-like, layered, platy	There are bedding, schistosity and joint, and often interlayer dislocation plane. The spacing between structural planes is 0.2 m - 0.4 m, and there are generally 3 groups of structural planes	As nearly homogeneous anisotropic bodies, the deformation and strength characteristics are controlled by the combination of bedding and strata, and they may be regarded as elastic-plastic bodies with poor stability	Potential collapse of unstable structural body, especially the buckling failure of rock strata and the plastic deformation of weak rock strata
Crushed stone-like and brecciated structure	Fractured rock strata affected severely by structure	Crushed stone-like and brecciated	Faults, fault fracture zones, schistosity, bedding and interlayer structural planes are well developed. The spacing between structural planes is less than 0.2 m and there are generally more than 3 groups of structural planes, which are composed of many detached bodies	The integrity is damaged greatly, the overall strength is low and controlled by structural planes such as fracture. The rock mass is mostly in the form of elastic-plastic medium with poor stability	It is easy to cause large-scale rock mass instability, and groundwater aggravates rock mass instability

Table 3.2.1(continued)

Name	Geologic body type	Shape of main structural body	Development of structural plane	Engineering characteristics of rock mass	Potential geotechnical problems
Loose structure	Rock mass in fault fracture zones affected severely by structure, in highly-weathered zones and completely-weathered zones	Fragmental or granular	The fault fracture zones cross, the structural and weathered fissures are dense, the structural planes and their combination are complicated and often filled with clay, forming many detached rock masses of different sizes	The integrity is damaged greatly, the stability is very poor and the rock mass property is close to loose medium	It is easy to cause large-scale rock mass instability, and groundwater aggravates rock mass instability

3.2.2 The classification of rock mass by the stratigraphic thickness of rock strata shall comply with Table 3.2.2.

Table 3.2.2 Classification of Rock Mass by Stratigraphic Thickness of Rock Strata

Name	Very thick layer	Thick layer	Medium-thick layer	Thin layer		
				Thin layer	Medium-thin layer	Very thin layer
Layer thickness h (m)	$h>1.0$	$0.5<h\leqslant 1.0$	$0.1<h\leqslant 0.5$	$0.03<h\leqslant 0.1$	$0.01<h\leqslant 0.03$	$h\leqslant 0.01$

3.2.3 The classification of rock mass by the joint width shall comply with Table 3.2.3.

Table 3.2.3 Classification of Rock Mass by Joint Width

Name	Joint width b (mm)
Closed joint	$b<1$
Slightly-open joint	$1\leqslant b<3$
Open joint	$3\leqslant b<5$
Widely-open joint	$b\geqslant 5$

3.2.4 The joint development degree of rock mass shall be determined according to Table 3.2.4.

Table 3.2.4 Classification of Joint Development Degree of Rock Mass

Joint development degree	Basic characteristics
Undeveloped	1-2 sets of joints, regular, structural type, with a spacing more than 1 m, mostly closed joints, with the rock mass being cut into huge blocks
Moderately developed	2-3 sets of joints, X-shaped, relatively regular, mostly of structural type, with a spacing more than 0.4 m mostly, mostly closed joints and partially slightly-open joints, with less filling materials, and with the rock mass being cut into massive blocks
Developed	More than 3 sets of joints, irregular, X-shaped or star-shaped, mainly of structural or weathered type, mostly with a spacing less than 0.4 m, mostly open joints, partially with filling materials and with the rock mass being cut into blocks
Well-developed	More than 3 sets of joints, disorderly, mainly of weathered or structural type, mostly with a spacing less than 0.2 m, mostly open joints, with individual widely-open joints, generally with filling materials, and with the rock mass being cut into fragments

3.2.5 The classification of rock mass by the degree of influence of geological structure shall comply with Table 3.2.5.

Table 3.2.5 Classification of Rock Mass by Degree of Influence of Geological Structure

Name	Basic characteristics
Slight	With small tectonic variation and undeveloped joints
Great	With large tectonic variation, located in the sections adjacent to faults or fold axis, with possible small faults and moderately-developed joints
Serious	With severe tectonic variation, located at the fold axis or in the sections affected by faults, with distortion and dragging of most soft rocks, and with developed joints
Very serious	Located in the fault fracture zone, with rock mass in the shape of block stone, crushed stone and angular gravel, sometimes in powder or soil, with well-developed joints

3.2.6 The classification of rock mass by integrity shall comply with Table 3.2.6.

Table 3.2.6 Classification of Rock Mass by Integrity

Name	Characteristics of structural plane	Structural type	Index of rock mass integrity K_v
Integrated	1 - 2 sets of structural planes, mainly of structural joints or layers, closed	Integral massive structure	$K_v > 0.75$
Moderately integrated	2 - 3 sets of structural planes, mainly of structural joints and layers, with fissures mostly of closed type and partially of slightly-open type, with less filling materials	Blocky structure	$0.55 < K_v \leqslant 0.75$
Moderately crushed	3 sets of structural planes generally, mainly of joints and weathered fissures, greatly affected by tectonism near faults, with fissures mostly of slightly-open type and open type, with many filling materials	Stratified structure, block stone and crushed stone-like structure	$0.35 < K_v \leqslant 0.55$
Crushed	More than 3 sets of structural planes, mainly of weathered fissures, greatly affected by tectonism near faults, with fissures mostly of open type, with many filling materials	Crushed stone-like and brecciated structure	$0.15 < K_v \leqslant 0.35$
Very crushed	With disorderly structural plane, greatly affected by tectonism near faults, with the widely-open fissures being filled with mud or mud intercalated with rock debris, with thick filling materials	Loose structure	$K_v \leqslant 0.15$

Notes: 1 For the type of the fissure width listed in the table, refer to Table 3.2.3 of this Standard;
2 In the table, the index of rock mass integrity is the square of the ratio of the elastic longitudinal wave velocity of rock mass to that of the intact rock in the same rock mass.

3.2.7 The zoning of rock mass by the weathering degree shall comply with Table 3.2.7.

Table 3.2.7 Weathering Degree Zoning of Rock Mass

Weathering degree zoning	Field identification characteristics				Parameter and index of weathering degree		
	Color of rocks and minerals	Structure	Degree of fragmentation	Hardness	Weathering coefficient K_f	Wave velocity ratio K_p	Longitudinal wave velocity v_p (m/s)
Unweathered	Rocks, minerals and their cements are in fresh colors and preserved in their original colors	The original structure of the rock mass is preserved	Except for the structural fissures, other fissures are invisible to naked eyes and the integrity is good	Except for the argillaceous rock that can be crushed with a sledge hammer, other rocks are not easy to break and only can be excavated by blasting	$K_f > 0.9$	$K_p > 0.9$	Hard rock $v_p > 5\,000$; Soft rock $v_p > 4\,000$

Table 3.2.7(continued)

Weathering degree zoning	Field identification characteristics				Parameter and index of weathering degree		
	Color of rocks and minerals	Structure	Degree of fragmentation	Hardness	Weathering coefficient K_f	Wave velocity ratio K_p	Longitudinal wave velocity v_p(m/s)
Slightly-weathered	Rocks and minerals are in dark colors and some minerals near the joint surface change colors	The rock mass structure is not damaged and weathering or incrustation exists only along the joint surface	There is a few weathering fissures mostly with a spacing more than 0.4 m, and the integrity is still good	A sledge hammer and wedge shall be used for splitting and the argillaceous rock may be crushed with a sledge hammer; excavation shall be conducted by blasting	Hard rock $0.8 < K_f \leqslant 0.9$; Soft rock $0.8 < K_f \leqslant 0.9$	Hard rock $0.8 < K_p \leqslant 0.9$; Soft rock $0.8 < K_p \leqslant 0.9$	Hard rock $4\,000 < v_p \leqslant 5\,000$; Soft rock $3\,000 < v_p \leqslant 4\,000$
Moderately-weathered	Rocks and minerals lose luster and are in dark colors; some easily-weathered minerals have been discolored and the biotite loses elasticity	The rock mass structure has been damaged partially and the fissures may be intercalated with weathered layers and are generally in blocky or spheroidal structure	The weathering fissures are developed mostly with a spacing of 0.2 m - 0.4 m, and the integrity is poor	Rocks can be crushed with a sledge hammer and are not easy to break with a hand hammer; most rocks need to be excavated by blasting and after that rocks may be drilled by a core drill	Hard rock $0.4 < K_f \leqslant 0.8$; Soft rock $0.3 < K_f \leqslant 0.8$	Hard rock $0.6 < K_p \leqslant 0.8$; Soft rock $0.5 < K_p \leqslant 0.8$	Hard rock $2\,000 < v_p \leqslant 4\,000$; Soft rock $1\,500 < v_p \leqslant 3\,000$
Highly-weathered	Rocks and most minerals change colors and the secondary minerals are formed	The rock mass structure has been damaged mostly and are in fragmental block structure or spheroidal structure	The weathering fissures are well developed, rocks are fragmental and the weathered materials are in gravel shape or sand-containing gravel shape; the spacing of fissures is less than 0.2 m and the integrity is poor	Rocks can be crushed with a hand hammer and excavated with a pickax, but it is difficult to use a shovel for excavation; they can be drilled by dry drilling	Hard rock $K_f \leqslant 0.4$; Soft rock $K_f \leqslant 0.3$	Hard rock $0.4 < K_p \leqslant 0.6$; Soft rock $0.3 < K_p \leqslant 0.5$	Hard rock $1\,000 < v_p \leqslant 2\,000$; Soft rock $700 < v_p \leqslant 1\,500$

Table 3.2.7 (continued)

Weathering degree zoning	Field identification characteristics				Parameter and index of weathering degree		
	Color of rocks and minerals	Structure	Degree of fragmentation	Hardness	Weathering coefficient K_f	Wave velocity ratio K_p	Longitudinal wave velocity v_p (m/s)
Completely-weathered	Rocks and minerals have been discolored completely and most of them mutate; except for quartz, most minerals have been weathered into soil	The rock mass structure has been damaged completely and the characteristics of the original rock are preserved only in appearance; the mineral crystals have lost connection and the quartz is loose and granular	weathered and fragmented, soil or sand	Rocks and minerals can be crumbed with hands and excavated with a shovel, and are easy to drill by dry drilling	—	Hard rock $K_p \leqslant 0.4$; Soft rock $K_p \leqslant 0.3$	Hard rock $500 < v_p \leqslant 1\,000$; Soft rock $300 < v_p \leqslant 700$

Notes: 1 K_f is the ratio of the saturated uniaxial compressive strength of weathered rock to that of unweathered rock in the same rock mass;

    2 K_p is the ratio of the longitudinal wave velocity of weathered rock to that of unweathered rock in the same rock mass.

3.2.8 The determination of the basic quality index of rock mass shall comply with the following requirements:

1 The basic quality index BQ of rock mass shall be calculated by Formula (3.2.8) according to the quantitative indexes of rock hardness and rock mass integrity.

$$BQ = 100 + 3R_c + 250K_v \qquad (3.2.8)$$

2 The following restrictions shall be followed when applying Formula (3.2.8):

    1) When R_c is greater than $90K_v + 30$, the BQ value shall be calculated by substituting $R_c = 90K_v + 30$ and K_v into the formula;

    2) When K_v is greater than $0.04R_c + 0.4$, the BQ value shall be calculated by substituting $K_v = 0.04R_c + 0.4$ and R_c into the formula.

3.2.9 The grading of basic quality of rock mass shall be determined from Table 3.2.9, based on the combination of qualitative characteristics of basic quality of rock mass and basic quality index BQ of rock mass.

Table 3.2.9 Grading of Basic Quality of Rock Mass

Grade of basic quality of rock mass	Qualitative characteristics of basic quality of rock mass	Basic quality index of rock mass (BQ)
I	Very hard rock, with integrated rock mass	>550
II	Very hard rock, with relatively integrated rock mass; Hard rock, with integrated rock mass	550 - 451
III	Very hard rock, with moderately crushed rock mass; Hard rock, with relatively integrated rock mass; Moderately soft rock, with integrated rock mass	450 - 351

Table 3.2.9(continued)

Grade of basic quality of rock mass	Qualitative characteristics of basic quality of rock mass	Basic quality index of rock mass (BQ)
IV	Very hard rock, with crushed rock mass; Hard rock, with moderately crushed - crushed rock mass; Moderately soft rock, with relatively integrated - moderately crushed rock mass; Soft rock, with integrated - relatively integrated rock mass	350 - 251
V	Moderately soft rock, with crushed rock mass; Soft rock, with moderately crushed - crushed rock mass; All very soft rocks and all very crushed rocks	≤250

Note: If the grade determined based on the qualitative characteristics of basic quality is inconsistent with that based on the basic quality index BQ of rock mass, the basic quality grade of rock mass shall be determined through qualitative classification and comprehensive analysis of quantitative indexes. If the grade difference is no less than one level, supplementary tests shall be conducted.

3.2.10 For preliminary classification of rock mass in railway tunnel engineering, slope engineering and ground engineering, the basic quality grade of rock mass determined according to Article 3.2.9 of this Standard may be regarded as the rock mass grade. For detailed classification, the corrected values of basic quality indexes of various engineering rock masses shall be determined according to the correction factors from groundwater, occurrence of structural plane, etc.

4 Classification of Soil

4.1 General Requirements

4.1.1 Soil shall be classified as follows according to the accumulation age, geological genesis, shape of soil particles, gradation or plasticity index.

1 It may be classified into early deposited soil (soil deposited in Q_3 and before), general deposited soil (soil deposited in Q_4^1) and recently deposited soil (soil deposited in Q_4^2) according to the accumulation age.

2 It may be classified into residual soil, slope wash, colluvial soil, diluvial soil, alluvial soil, marine soil, lacustrine soil, morainic soil, glacial deposition soil and eolian soil according to the geological genesis.

3 It may be classified into gravel, sand, silt and clay according to the shape of soil particles, gradation or plasticity index.

4.1.2 The soil layer rhythmically deposited should be named as interlayer when the ratio of thin layer thickness to thick layer thickness is 1/10 - 1/3, and the thick layer shall be written in the front; when the thickness ratio is greater than 1/3, it should be named as interbedding; when the thickness ratio is less than 1/10, it should be named as laminated thin layer.

4.1.3 The soil composed of coarse and fine particles of discontinuous particle gradation formed from slope wash, diluvial soil and glacial deposition soil shall be named as mixed soil, and the soil name shall be preceded by the name of the main substance. When the mass of the main substance accounts for 5% - 25% of the total mass, it shall be understood that the content of substances is slight; when the mass content of the substances is greater than or equal to 25%, it shall be understood that such substance are one of the main content.

4.2 Classification of General Soil

4.2.1 The grouping of soil particles shall comply with Table 4.2.1.

Table 4.2.1 Particle Size Grouping of Soil

Particle name		Particle size d (mm)
Boulder (rounded or subrounded) or rubble (sharp angular)	Large	$d > 800$
	Medium	$400 < d \leqslant 800$
	Small	$200 < d \leqslant 400$
Pebble (rounded or subrounded) or gravel (sharp angular)	Large	$100 < d \leqslant 200$
	Small	$60 < d \leqslant 100$
Coarse round gravel (rounded or subrounded) or coarse angular gravel (sharp angular)	Large	$40 < d \leqslant 60$
	Small	$20 < d \leqslant 40$
Fine round gravel (rounded or subrounded) or fine angular gravel (sharp angular)	Large	$10 < d \leqslant 20$
	Medium	$5 < d \leqslant 10$
	Small	$2 < d \leqslant 5$

Table 4.2.1(continued)

Particle name		Particle size d (mm)
Sandy particle	Coarse	$0.5 < d \leqslant 2$
	Medium	$0.25 < d \leqslant 0.5$
	Fine	$0.075 < d \leqslant 0.25$
Silty particle		$0.005 < d \leqslant 0.075$
Clayey particle		$d < 0.005$

4.2.2 The classification of gravel shall comply with the following requirements:

1 The classification based on the shape and gradation of soil particle shall comply with Table 4.2.2-1.

Table 4.2.2-1 Classification of Gravel

Soil name	Particle shape	Particle size gradation
Bouldery soil	Most are rounded or subrounded	Mass percentage of particles with a particle size being greater than 200 mm exceeds 50% of total mass
Subangular-boulder soil	Most are sharp angular	
Cobble soil	Most are rounded or subrounded	Mass percentage of particles with a particle size being greater than 60 mm exceeds 50% of total mass
Gravelly soil	Most are sharp angular	
Coarse round gravel	Most are rounded or subrounded	Mass percentage of particles with a particle size being greater than 20 mm exceeds 50% of total mass
Coarse angular gravel	Most are sharp angular	
Fine round gravel	Most are rounded or subrounded	Mass percentage of particles with a particle size being greater than 2 mm exceeds 50% of total mass
Fine angular gravel	Most are sharp angular	

Note: The soil name shall be determined according to the particle size grouping (from large to small) and subject to the first conforming item.

2 The qualitative description of compactness may be determined as per Table 4.2.2-2 according to the structural characteristics, geomorphic features, natural slope form, excavation and drilling conditions.

Table 4.2.2-2 Classification of Compactness for Gravel

Compactness	Structural characteristics	Natural slope and excavation conditions	Drilling conditions
Dense	The skeleton particles are interlaced and in close and continuous contact with each other; the pores are filled up densely	The natural steep slope is stable with less deposits under the slope. It is difficult to dig with a pickax and the slope can be loosened only with crowbars. The pit wall is stable and the part of the pit wall where large particles are taken out can be kept in a concave shape	Very difficult for drilling. During drilling, the drilling tool jumps severely and the hole wall is stable
Medium dense	The skeleton particles are unevenly arranged and some particles are not in contact with each other; the pores are filled up but not compacted	The natural steep slope is not easy to rise or there are much deposits under the slope. The inclination of the natural slope is greater than the angle of repose of coarse particles. It can be excavated with a pickax and the pit wall is subject to spalling. When it is filled with sandy soil, the part of the pit wall where large particles are taken out are difficult to be kept in a concave shape	Difficult for drilling. During drilling, the drilling tool does not jump severely and the hole wall collapses

Table 4.2.2-2(continued)

Compactness	Structural characteristics	Natural slope and excavation conditions	Drilling conditions
Slightly dense	Most of the skeleton particles are not in contact with each other and the pores are basically filled up but loosely	A steep slope is difficult to form and the natural slope is slightly greater than the angle of repose of coarse particles. It can be easily excavated with a pickax and the pit wall is prone to spalling. The part of the pit wall where large particles are taken out is prone to collapse	Difficult for drilling. During drilling, the drilling tool jumps and the hole wall is easy to collapse
Loose	The skeleton particles have large pores filled insufficiently and loosely	It can be excavated with a shovel. The natural slope is mainly of the angle of repose of main particles. The pit wall is prone to collapse	Easy for drilling. The hole wall is easy to collapse during drilling

3 For the gravel with an average particle size being less than or equal to 50 mm and the maximum particle size being less than 100 mm, the compactness shall be evaluated quantitatively according to Table 4.2.2-3.

Table 4.2.2-3 Classification of Compactness of Gravel Based on $N_{63.5}$

Measured blow counts of heavy dynamic penetration test ($N_{63.5}$)	Compactness	Measured blow counts of heavy dynamic penetration test ($N_{63.5}$)	Compactness
$N_{63.5} \leqslant 5$	Loose	$10 < N_{63.5} \leqslant 20$	Medium dense
$5 < N_{63.5} \leqslant 10$	Slightly dense	$N_{63.5} > 20$	Dense

4 For the gravel with an average particle size being greater than 50 mm or the maximum particle size being greater than 100 mm, the compactness shall be evaluated quantitatively according to Table 4.2.2-4.

Table 4.2.2-4 Classification of Compactness of Gravel Based on N_{120}

Measured blow counts of extra-heavy dynamic penetration test (N_{120})	Compactness	Measured blow counts of extra-heavy dynamic penetration test (N_{120})	Compactness
$N_{120} \leqslant 3$	Loose	$11 < N_{120} \leqslant 14$	Dense
$3 < N_{120} \leqslant 6$	Slightly dense	$N_{120} > 14$	Very dense
$6 < N_{120} \leqslant 11$	Medium dense		

5 The humidity shall be classified as per Table 4.2.2-5 according to the degree of saturation. The degree of saturation S_r shall be calculated by the following Formula:

$$S_r = \frac{V_w}{V_v} \times 100\% \tag{4.2.2}$$

Where

V_w—water volume;

V_v—volume of pores (including water and air).

Table 4.2.2-5 Classification of Humidity for Gravel

Classification	Degree of Saturation S_r (%)
Slightly wet	$S_r \leqslant 50$
Wet	$50 < S_r \leqslant 80$
Saturated	$S_r > 80$

4.2.3 The classification of sands shall comply with the following requirements.

1 The classification based on the particle gradation shall comply with Table 4.2.3-1.

Table 4.2.3-1 Classification of Sands

Soil name	Particle size gradation
Gravelly sand	Mass percentage of particles with a particle size being greater than 2 mm accounts for 25% - 50% of total mass
Coarse sand	Mass percentage of particles with a particle size being greater than 0.5 mm accounts for 50% of total mass
Medium sand	Mass percentage of particles with a particle size being greater than 0.25 mm accounts for 50% of total mass
Fine sand	Mass percentage of particles with a particle size being greater than 0.075 mm accounts for 85% of total mass
Silty sand	Mass percentage of particles with a particle size being greater than 0.075 mm accounts for 50% of total mass

Note: The name shall be determined according to the particle size gradation (from large to small) and subject to the first conforming item.

2 The compactness shall be classified as per Table 4.2.3-2 according to the measured blow counts of standard penetration test or relative density. The relative density D_r shall be calculated by the following Formula:

$$D_r = \frac{e_{max} - e}{e_{max} - e_{min}} \tag{4.2.3}$$

Where

e — natural void ratio;

e_{max} — maximum void ratio;

e_{min} — minimum void ratio.

Table 4.2.3-2 Classification of Compactness for Sands

Compactness	Measured blow counts N of standard penetration test	Relative density D_r
Dense	$N > 30$	$D_r > 0.67$
Medium dense	$15 < N \leq 30$	$0.4 < D_r \leq 0.67$
Slightly dense	$10 < N \leq 15$	$0.33 < D_r \leq 0.4$
Loose	$N \leq 10$	$D_r \leq 0.33$

3 The humidity shall be classified as per Table 4.2.3-3 according to the degree of saturation. The degree of saturation S_r shall be calculated by Formula (4.2.2).

Table 4.2.3-3 Classification of Humidity for Sands

Classification	Degree of Saturation S_r (%)
Slightly wet	$S_r \leq 50$
Wet	$50 < S_r \leq 80$
Saturated	$S_r > 80$

4.2.4 The soil whose plasticity index is equal to or less than 10 and whose mass percentage of particles with a particle size greater than 0.075 mm does not exceed 50% of the total mass shall be named as silt.

1 The compactness of silt shall be classified as per Table 4.2.4-1 according to the void ratio.

Table 4.2.4-1 Classification of Compactness of Silt

Compactness	Void ratio e
Dense	$e < 0.75$
Medium dense	$0.75 \leq e \leq 0.9$
Slightly dense	$e > 0.9$

2 The humidity of silt shall be classified as per Table 4.2.4-2 according to the natural moisture content.

Table 4.2.4-2 Classification of Humidity for Silt

Classification	Natural water ratio w (%)
Slightly wet	$w<20$
Wet	$20 \leqslant w \leqslant 30$
Saturated	$w>30$

4.2.5 The soil whose plasticity index is greater than 10 shall be named as clay.

1 The clay shall be classified as per Table 4.2.5-1 according to the plasticity index.

Table 4.2.5-1 Classification of Clay

Soil name	Plasticity index I_P
Silty clay	$10<I_P \leqslant 17$
Clay	$I_P>17$

Note: The plasticity index I_P is the difference between the moisture content at liquid limit and that at plastic limit of soil. The liquid and plastic limits shall be determined by the liquid-plastic limit combined measurement method, and the liquid limit is at 10 mm.

2 The compressibility of clay shall be classified according to Table 4.2.5-2.

Table 4.2.5-2 Classification of Compressibility for Clay

Compressibility grade		Coefficient of compressibility $a_{0.1-0.2}$ (MPa^{-1})
Low compressibility		$a_{0.1-0.2}<0.1$
Medium compressibility	Medium-low compressibility	$0.1 \leqslant a_{0.1-0.2}<0.3$
	Medium-high compressibility	$0.3 \leqslant a_{0.1-0.2}<0.5$
High compressibility		$a_{0.1-0.2} \geqslant 0.5$

Note: $a_{0.1-0.2}$ is the coefficient of compressibility within the pressure ranging from 0.1 MPa to 0.2 MPa.

3 The plasticity of clay shall be classified according to Table 4.2.5-3. The liquidity index I_L shall be calculated by the following Formula:

$$I_L = \frac{w - w_P}{I_P} \qquad (4.2.5)$$

Where

w—natural moisture content;

w_P—moisture content at plastic limit;

I_P—plasticity index.

Table 4.2.5-3 Classification of Plasticity for Clay

Plasticity	Liquidity index I_L
Stiff	$I_L \leqslant 0$
Stiff-plastic	$0<I_L \leqslant 0.50$
Soft-plastic	$0.50<I_L \leqslant 1$
Liquid-plastic	$I_L>1$

4.3 Classification of Special Soil

4.3.1 The special soil may be classified into loess, red clay, swelling soil, soft soil, saline soil, permafrost, seasonally frozen soil and fill according to the content of special substances, structural characteristics and special engineering geological properties of soil.

4.3.2 The judgment and classification of loess shall comply with the following requirements:

1 The soil formed under arid and semi-arid climate conditions since Quaternary period, with the soil particles being dominated by silt and containing calcium carbonate and a small amount of soluble salt, and with the engineering geological characteristics of large pores and vertical joints, poor water resistance, prone to disintegration and suffosion, and collapsibility of upper part.

2 The classification of loess by the accumulation age shall comply with Table 4.3.2-1.

Table 4.3.2-1 Classification of Loess by Accumulation Age

Stratigraphic age		Stratum name	Collapsibility and other characteristics
Holocene epoch Q_4	Recent Q_4^2	New loess	Recently deposited loess-like soil	Self-weight or non-self-weight collapsible loess, with high compressibility
	Early Q_4^1		Loess-like soil	
Late Pleistocene epoch Q_3			Malan loess	
Middle Pleistocene epoch Q_2		Old loess	Lishi loess	With collapsibility of upper soil of some soil layers
Early Pleistocene epoch Q_1			Wucheng loess	No collapsibility

Note: The collapsibility of loess below the top surface of Q_2 Lishi loess shall be determined through laboratory collapsibility tests or field immersion tests according to the actual pressure of building or the saturated self-weight pressure of overburden soil.

3 The classification of loess by the plasticity index I_P shall comply with Table 4.3.2-2.

Table 4.3.2-2 Classification of Loess by Plasticity Index

Name	Plasticity index I_P
Sandy loess	$I_P \leqslant 10$
Clayey loess	$I_P > 10$

4 The collapsibility of loess shall be determined as per Table 4.3.2-3 according to the collapsibility coefficient measured by the laboratory collapsibility test under certain pressure.

Table 4.3.2-3 Classification of Collapsibility of Loess

Name	Collapsibility coefficient δ_S
Non-collapsible loess	$\delta_S < 0.015$
Collapsible loess	$\delta_S \geqslant 0.015$

5 The loess shall be named as self-weight collapsible loess if it is collapsible when immersed in water under the gravitational pressure of overburden soil; the loess shall be named as non-self-weight collapsible loess if it is collapsible when immersed in water under the pressure being greater than the gravitational pressure (including the gravity of soil and additional pressure) of overburden soil.

4.3.3 The judgment and classification of red clay shall comply with the following requirements:

1 The primary red clay is brownish red or brownish yellow and overlies the carbonate rock strata, with a liquid limit being equal to or greater than 50% and with high plasticity. The primary red clay shall be judged as secondary red clay if the basic characteristics of residual clay are retained after transportation and deposition, and the liquid limit is greater than 45%. The engineering geological characteristics of red clay include softening in water, strong shrinkage under water loss, fissure development and prone to spalling.

2 The classification of plasticity for red clay shall comply with Table 4.3.3-1.

Table 4.3.3-1 Classification of Plasticity for Red Clay

State	Water content ratio a_w	Specific penetration resistance P_s (MPa)	Empirical reference
Stiff	$a_w \leqslant 0.55$	$P_s \geqslant 2.3$	Dry and hard
Stiff-plastic	$0.55 < a_w \leqslant 0.70$	$1.3 \leqslant P_s < 2.3$	Not easily rubbed into 3 mm-thick soil strips
Soft-plastic	$0.70 < a_w \leqslant 1.00$	$0.2 \leqslant P_s < 1.3$	Easily rubbed into 3 mm-thick soil strips
Liquid-plastic	$a_w > 1.00$	$P_s < 0.2$	Wet, near or in a flow state

Note: The water content ratio is the ratio of the natural moisture content to the liquid limit of soil. The liquid limit shall be determined by the liquid-plastic limit combined measurement method, and the liquid limit is at 10 mm.

3 The classification of fissure state for red clay shall comply with Table 4.3.3-2.

Table 4.3.3-2 Classification of Fissure State for Red Clay

Fissure state	Appearance features
Compact	Occasional fissures, less than 1/m
Massive	Many fissures, 1/m - 5/m
Fragmental	fissures well, more than 5/m

4 The classification of red clay based on the liquidity-plasticity limit ratio and the liquidity-plasticity ratio relationship shall comply with Table 4.3.3-3. The liquidity-plasticity limit ratio I'_r and liquidity-plasticity ratio I_r shall be calculated by the following formulae:

$$I'_r = 1.4 + 0.006 w_L \quad (4.3.3\text{-}1)$$

$$I_r = \frac{w_L}{w_p} \quad (4.3.3\text{-}2)$$

Where

w_L—liquid limit of soil;

w_p—plastic limit of soil.

Table 4.3.3-3 Classification of Red Clay Based on Liquidity-Plasticity Limit Ratio and Liquidity-Plasticity Ratio Relationship

Type	Relationship between I_r and I'_r	Shrinkage characteristics
Type I	$I_r \geqslant I'_r$	The swelling amount can be restored to original value after shrinkage and re-immersion
Type II	$I_r < I'_r$	The swelling amount can not be restored to original value after shrinkage and re-immersion

4.3.4 The judgment and classification of swelling soil shall comply with the following requirements:

1 Swelling soil is a cohesive soil with high liquid limit which is mainly composed of montmorillonite, illite and other hydrophilic minerals. Its properties change with the environment humidity. It significantly swells, softens and disintegrates upon absorption of water, and shrink abruptly, cracks and hardens due to water loss. It is characterized by cyclic swelling-shrinkage deformation.

2 The swelling soil shall be judged preliminarily and in detail according to the following requirements.

1) The swelling soil shall be judged preliminarily as per Table 4.3.4-1 according to the topographic features, soil color, structure, soil texture, natural geological phenomenon, free swelling ratio of soil and other characteristics.

Table 4.3.4-1   Conditions for Preliminary Judgment of Swelling Soil

Topographic features	Ridge-type geomorphologic landscape, ridges alternating with valleys; gentle and open terrain, without natural scarp and slope groove is developed
Color	Mostly in brown, yellow and dark-brown, intercalated with grayish white or grayish green strips or film; grayish white or grayish green strips or film in the form of lens or interlayers
Structure	Multi-fissure structure with irregular direction. The fissure surface is smooth with common scratches. Fissures are often filled with grayish white or grayish green clay
Texture	With fine texture and slippery feeling when touching the soil, it often contains calcareous or ferrimanganic nodules or peastones, which are locally enriched into layers
Natural geological phenomenon	Shallow sliding, landslide or ground cracking can be usually found on the slope surface. If the soil slope is layered, the swelling soil tends to form a concave slope. Newly-excavated pit walls are prone to collapse
Free swelling ratio F_S (%)	$F_S \geqslant 40$

  2) Three indexes including the free swelling ratio, montmorillonite content and cation exchange capacity shall be adopted for the detailed judgment of swelling soil. When any two or more indexes in Table 4.3.4-2 are satisfied, the soil shall be judged as swelling soil.

Table 4.3.4-2   Detailed Judgment Indexes of Swelling Soil

Name	Index
Free swelling ratio F_S (%)	$F_S \geqslant 40$
Montmorillonite content M (%)	$M \geqslant 7$
Cation exchange capacity CEC(NH_4^+) (mmol/kg)	CEC(NH_4^+) $\geqslant 170$

Note: CEC(NH_4^+) represents the exchange capacity of cations (NH_4^+) in 1 kg dry soil.

3 The swelling potential of swelling soil shall be graded according to Table 4.3.4-3.

Table 4.3.4-3   Grading of Swelling Potential of Swelling Soil

Index	Grade		
	Low swelling soil	Moderate swelling soil	High swelling soil
Free swelling ratio F_S (%)	$40 \leqslant F_S < 60$	$60 \leqslant F_S < 90$	$F_S \geqslant 90$
Montmorillonite content M (%)	$7 \leqslant M < 17$	$17 \leqslant M < 27$	$M \geqslant 27$
Cation exchange capacity CEC(NH_4^+) (mmol/kg)	$170 \leqslant$ CEC(NH_4^+) < 260	$260 \leqslant$ CEC(NH_4^+) < 360	CEC(NH_4^+) $\geqslant 360$

Note: When any two or more indexes in a column are satisfied, the swelling property shall be determined as the corresponding grade.

4.3.5 The judgment and classification of soft soil shall comply with the following requirements:

  1 Soft soil is the cohesive soil with a natural void ratio greater than or equal to 1.0, a natural moisture content greater than or equal to the liquid limit, a coefficient of compressibility greater than or equal to 0.5 MPa^{-1}, and undrained shear strength less than 30 kPa. The soft soil usually contains organic matters and has the characteristics of high compressibility, low strength and high sensitivity, and deforms slowly in drained consolidation test.

  2 The classification of soft soil based on the physical and mechanical properties may be analyzed and determined according to Table 4.3.5-1.

Table 4.3.5-1 Classification of Soft Soil

Classification index		Name				
		Soft clay	Muddy soil	Mud	Peat soil	Peat
Organic matter content W_u	%	$W_u<3$	$3 \leqslant W_u<10$		$10 \leqslant W_u \leqslant 60$	$W_u>60$
Natural void ratio e	—	$e \geqslant 1.0$	$1.0 \leqslant e \leqslant 1.5$	$e>1.5$	$e>3$	$e>10$
Natural moisture content w	%	$w \geqslant w_L$			$w \gg w_L$	
Permeability coefficient k	cm/s	$k<10^{-6}$			$k<10^{-3}$	$k<10^{-2}$
Coefficient of compressibility $a_{0.1-0.2}$	MPa^{-1}	$a_{0.1-0.2} \geqslant 0.5$			—	
Undrained shear strength C_u	kPa	$C_u<30$			$C_u<10$	
Specific penetration resistance of static cone penetration test P_s	kPa	$P_s<700$				
Cone tip resistance of static cone penetration test q_c	kPa	$q_c<600$				
Number of blows in standard penetration test N	Blows	$N<4$			$N<2$	
Vane shear strength S_u	kPa	$\mu S_u<30$				
Soil type index I_D	—	$I_D<0.35$				

Notes: 1 Silt can be named as "soft silt" when its physical and mechanical properties are mostly consistent with the indexes in the table.
2 μ is the correction coefficient: $I_p \leqslant 20$, $\mu=1$; $20<I_p \leqslant 40$, $\mu=0.9$.

3 The classification of soft soil by its genesis shall comply with Table 4.3.5-2.

Table 4.3.5-2 Genetic Type of Soft Soil

Topographic features	Genetic type	Sedimentary characteristics
Coastal plain	Littoral facies	Uneven and very loose formation, often mixed with sand and gravel layers
	Littoral - neritic facies	Uneven and loose formation, often mixed with sand and silty soil
	Lagoon facies	Fine particles, large void ratio, low strength, often intercalated with thin layer of peat
	Liman facies	Large void ratio, loose structure and high moisture content
	Delta facies	Poor sorting, loose structure, multiple cross-bedding, and many thin layers of silty sand
Lacustrine plain	Lacustrine facies	High content of silt particles, with obvious bedding and soft structure; irregular thickness of surface crust
River alluvial plain	Flood plain facies Banco facies	Complex stratification with heterogeneous composition, mainly composed of mud and soft clay, interbedded with sand or peat
Mountainous valley	Valley facies	The soft soil is distributed in schistose or banded form, shallow near mountain side and deep at the middle of valley, with a great change in thickness. The particles become fine from the front of the mountain to the middle of valley. The underlying hard bottom slope is steep
Peat bog	Swamp facies	It is mainly composed of peat and often exposed to the surface. The pores are large and elastic. The lower part has mud or thin layer of mud interbedded with peat

4 The sensitivity of soft soil shall be determined as per Table 4.3.5-3 through the unconfined compressive strength test or field vane shear test.

Table 4.3.5-3 Classification of Sensitivity for Soft Soil

Classification	Sensitivity S_t
Low sensitivity	$S_t \leqslant 2$
Medium sensitivity	$2 < S_t \leqslant 4$
High sensitivity	$4 < S_t \leqslant 8$
Extreme sensitivity	$8 < S_t \leqslant 16$
Fluidity	$S_t > 16$

1) The sensitivity S_t of soft soil shall be calculated by the following formula when unconfined compressive strength test is conducted:

$$S_t = \frac{q_u}{q'_u} \qquad (4.3.5\text{-}1)$$

Where

q_u—unconfined compressive strength of undisturbed soil (kPa);

q'_u—unconfined compressive strength of remoulded soil whose density and moisture content are the same as those of undisturbed soil but whose structure is completely destroyed.

2) The sensitivity S_t of soft soil shall be calculated by the following formula when field vane shear test is conducted:

$$S_t = \frac{S_u}{S'_u} \qquad (4.3.5\text{-}2)$$

Where

S_u—vane shear strength (kPa);

S'_u—remoulded vane shear strength (kPa).

4.3.6 The judgment and classification of saline soil shall comply with the following requirements:

1 Saline soil is the soil with soluble salt content greater than 0.3%. If the average content of soluble salt within 1.0 m below the ground surface is greater than 0.3%, the site concerned shall be classified as saline soil area or site. The saline soil has the engineering geological characteristics of high hygroscopicity, dilatation, collapsibility and corrosion.

2 The classification of saline soil by the salt nature shall comply with Table 4.3.6-1.

Table 4.3.6-1 Classification of Saline Soil by Salt Nature

Saline soil	Salt content ratio D_1	Salt content ratio D_2
Chloride saline soil	$D_1 > 2$	—
Subchloride saline soil	$2 \geqslant D_1 > 1$	—
Sulfurous acid saline soil	$1 \geqslant D_1 \geqslant 0.3$	—
Sulfate saline soil	$D_1 < 0.3$	—
Alkaline saline soil	—	$D_2 > 0.3$

3 The salt nature of saline soil shall be calculated by the following Formulae:

$$D_1 = \frac{b(\text{Cl}^-)}{2b(\text{SO}_4^{2-})} \qquad (4.3.6\text{-}1)$$

$$D_2 = \frac{2b(\text{CO}_3^{2-}) + b(\text{HCO}_3^-)}{b(\text{Cl}^-) + 2b(\text{SO}_4^{2-})} \qquad (4.3.6\text{-}2)$$

Where, $b(Cl^-)$, $b(HCO_3^{2-})$, $2b(SO_4^{2-})$ and $2b(CO_3^{2-})$ refer to the molality of the matter in brackets contained in 1 kg soil (mmol/kg).

4 The classification of saline soil by the salinity degree shall comply with Table 4.3.6-2.

Table 4.3.6-2 Classification of Saline Soil by Salinity Degree

Salinity degree	Average salt content of soil $\overline{DT}$ (%)		
	Chloride saline soil and subchloride saline soil	Sulfate saline soil and fulfite acid saline soil	Alkaline saline soil
Slightly saline soil	$0.3 < \overline{DT} \leqslant 1.0$	—	—
Moderately saline soil	$1.0 < \overline{DT} \leqslant 5.0$	$0.3 < \overline{DT} \leqslant 2.0$	$0.3 < \overline{DT} \leqslant 1.0$
Highly saline soil	$5.0 < \overline{DT} \leqslant 8.0$	$2.0 < \overline{DT} \leqslant 5.0$	$1.0 < \overline{DT} \leqslant 2.0$
Very highly saline soil	$\overline{DT} > 8.0$	$\overline{DT} > 5.0$	$\overline{DT} > 2.0$

Note: The "average salt content" in the table is calculated as a weighted average of the thickness represented by the sample.

4.3.7 Frozen soil is the soil (rock) containing ice crystals at a temperature of 0 ℃ or below. It may be classified into permafrost and seasonally frozen soil.

1 Permafrost is the soil that has been frozen for a period of two or more years. The physical and mechanical properties of permafrost change accordingly with the temperature, and frost heaving, thaw settlement or thaw slumping may occur.

2 Permafrost shall be classified as per Table 4.3.7-1 according to the soil type, total moisture content and average thaw subsidence coefficient δ_0.

The average thaw subsidence coefficient δ_0 shall be calculated by the following Formula:

$$\delta_0 = \frac{h_1 - h_2}{h_1} \times 100\% \tag{4.3.7-1}$$

Where
 h_1—height of permafrost samples before thawing (mm);
 h_2—height of permafrost samples after thawing (mm).

Table 4.3.7-1 Classification of Permafrost

Type of permafrost	Soil name	Total moisture content w_A (%)	Humidity after thawing	Average thaw subsidence coefficient δ_0 (%)	Thaw-settlement grade	Thaw-settlement type
Ice-poor permafrost	Gravel, gravelly sand, coarse sand and medium sand (with the mass of silt and clay being not greater than 15%)	$w_A < 10$	Wet	$\delta_0 \leqslant 1$	I	No thaw-settlement
	Gravel, gravelly sand, coarse sand and medium sand (with the mass of silt and clay being greater than 15%)	$w_A < 12$	Slightly wet			
	Fine sand and silty sand	$w_A < 14$				
	Silt	$w_A < 17$				
	Clay	$w_A < w_p$	Stiff			

Table 4.3.7-1(continued)

Type of permafrost	Soil name	Total moisture content w_A (%)	Humidity after thawing	Average thaw subsidence coefficient δ_0 (%)	Thaw-settlement grade	Thaw-settlement type
Ice-moderate permafrost	Gravel, gravelly sand, coarse sand and medium sand (with the mass of silt and clay being not greater than 15%)	$10 \leqslant w_A < 15$	Saturated	$1 < \delta_0 \leqslant 3$	II	Moderate thaw-settlement
	Gravel, gravelly sand, coarse sand and medium sand (with the mass of silt and clay being greater than 15%)	$12 \leqslant w_A < 15$	Wet			
	Fine sand and silty sand	$14 \leqslant w_A < 18$				
	Silt	$17 \leqslant w_A < 21$				
	Clay	$w_p \leqslant w_A < w_p + 4$	Stiff-plastic			
Ice-rich permafrost	Gravel, gravelly sand, coarse sand and medium sand (with the mass of silt and clay being not greater than 15%)	$15 \leqslant w_A < 25$	Saturated water yield (with the yield amount being less than 10%)	$3 < \delta_0 \leqslant 10$	III	Thaw-settlement
	Gravel, gravelly sand, coarse sand and medium sand (with the mass of silt and clay being greater than 15%)		Saturated			
	Fine sand and silty sand	$18 \leqslant w_A < 28$				
	Silt	$21 \leqslant w_A < 32$				
	Clay	$w_p + 4 \leqslant w_A < w_p + 15$	Soft-plastic			
Ice-saturated permafrost	Gravel, gravelly sand, coarse sand and medium sand (with the mass of silt and clay being less than 15%)	$25 \leqslant w_A < 44$	Saturated with water yield (with the yield amount being less than 10%)	$10 < \delta_0 \leqslant 25$	IV	High thaw-settlement
	Gravel, gravelly sand, coarse sand and medium sand (with the mass of silt and clay being greater than 15%)		Saturated			
	Fine sand and silty sand	$28 \leqslant w_A < 44$				
	Silt	$32 \leqslant w_A < 44$				
	Clay	$w_p + 15 \leqslant w_A < w_p + 35$	Soft-plastic			

Table 4.3.7-1(continued)

Type of permafrost	Soil name	Total moisture content w_A (%)	Humidity after thawing	Average thaw subsidence coefficient δ_0 (%)	Thaw-settlement grade	Thaw-settlement type
Ice layer with soil	Gravel, sandy soil and silt	$w_A \geqslant 44$	Saturated with large amount of water yield (with the yield amount of 10% - 20%)	$\delta_0 > 25$	V	Thawing collapse
	clay	$w_A \geqslant w_p + 35$	Liquid-plastic			
Pure ice layer	With a thickness of more than 25 cm or with the cumulative thickness of ice layers at an interval of 2 cm - 3 cm being more than 25 cm					

Notes: 1 The total moisture content shall contain ice and unfrozen water;
   2 The saline permafrost, peatificated permafrost, humus, and clay with high plasticity are not included in the table;
   3 w_p is the moisture content at plastic limit.

3 The seasonally thawed layer of permafrost is the crust surface that annually freezes in the cold season and thaws in the warm season.

4 The seasonally frozen soil is the soil (rock) in the crust surface that freezes in the cold season and thaws in the warm season.

5 The frost-heaving property of seasonally frozen soil and seasonally thawed layer of permafrost shall be classified as per Table 4.3.7-2 according to the average frost-heaving ratio of soil. The average frost-heaving ratio η of the frozen soil layer shall be calculated by the following formula:

$$\eta = \frac{\Delta_z}{h - \Delta_z} \times 100\% \quad\quad\quad (4.3.7\text{-}2)$$

Where

  Δ_z—surface frost-heaving amount (mm);

  h—thickness of frozen layer (mm).

Table 4.3.7-2 Frost-heaving Property Classification of Seasonally Frozen Soil and Seasonally Thawed Soil Layer

Soil type	Natural moisture content before freezing w (%)	Minimum distance from the groundwater level to the design frost depth before freezing h_w (m)	Average frost-heaving ratio η (%)	Frost-heaving grade	Frost-heaving type
Coarse-grained soil with the mass of silt and clay being not greater than 15% (including gravel, gravelly sand, coarse sand and medium sand, the same below) and fine sand with the mass of silt and clay being not greater than 10%	Unsaturated	Not considered	$\eta \leqslant 1$	Class I	No frost-heaving
Coarse-grained soil with the mass of silt and clay being greater than 15% and fine sand with the mass of silt and clay being greater than 10%	$w \leqslant 12$	>1.0			

Table 4.3.7-2(continued)

Soil type	Natural moisture content before freezing w (%)	Minimum distance from the groundwater level to the design frost depth before freezing h_w (m)	Average frost-heaving ratio η (%)	Frost-heaving grade	Frost-heaving type
Silty sand	$12 < w \leqslant 14$	>1.0	$\eta \leqslant 1$	Class I	No frost-heaving
Silt	$w \leqslant 19$	>1.5			
Clay	$w \leqslant w_p + 2$	>2.0			
Coarse-grained soil with the mass of silt and clay being not greater than 15% and fine sand with the mass of silt and clay being not greater than 10%	Saturated with moisture	No aquiclude	$1 < \eta \leqslant 3.5$	Class II	Moderate frost-heaving
Coarse-grained soil with the mass of silt and clay particles being greater than 15% and fine sand with the mass of silt and clay being greater than 10%	$w \leqslant 12$	$\leqslant 1.0$			
	$12 < w \leqslant 18$	>1.0			
Silty sand	$w \leqslant 14$	$\leqslant 1.0$			
	$14 < w \leqslant 19$	>1.0			
Silt	$w \leqslant 19$	$\leqslant 1.5$			
	$12 < w \leqslant 22$	>1.5			
Clay	$w \leqslant w_p + 2$	$\leqslant 2.0$			
	$w_p + 2 < w \leqslant w_p + 5$	>2.0			
Coarse-grained soil with the mass of silt and clay being not greater than 15% and fine sand with the mass of silt and clay particles being not greater than 10%	Saturated with moisture	With aquiclude	$3.5 < \eta \leqslant 6$	Class III	Frost-heaving
Coarse-grained soil with the mass of silt and clay being greater than 15% and fine sand with the mass of silt and clay particles being greater than 10%	$12 < w \leqslant 18$	$\leqslant 1.0$			
	$w > 18$	>0.5			
Silty sand	$14 < w \leqslant 19$	$\leqslant 1.0$			
	$19 < w \leqslant 23$	>1.0			
Silt	$19 < w \leqslant 22$	$\leqslant 1.5$			
	$22 < w \leqslant 26$	>1.5			
Clay	$w_p + 2 < w \leqslant w_p + 5$	$\leqslant 2.0$			
	$w_p + 5 < w \leqslant w_p + 9$	>2.0			

Table 4.3.7-2 (continued)

Soil type	Natural moisture content before freezing w (%)	Minimum distance from the groundwater level to the design frost depth before freezing h_w (m)	Average frost-heaving ratio η (%)	Frost-heaving grade	Frost-heaving type
Coarse-grained soil with the mass of silt and clay being greater than 15% and fine sand with the mass of silt and clay being greater than 10%	$w>18$	$\leqslant 0.5$	$6<\eta \leqslant 12$	Class IV	High frost-heaving
Silty sand	$19<w \leqslant 23$	$\leqslant 1.0$			
Silt	$22<w \leqslant 26$	$\leqslant 1.5$			
Silt	$26<w \leqslant 30$	>1.5			
Clay	$w_p+5<w \leqslant w_p+9$	$\leqslant 2.0$			
Clay	$w_p+9<w \leqslant w_p+15$	>2.0			
Silty sand	$w>23$	Not considered	$\eta>12$	Class V	Very high frost-heaving
Silt	$26<w \leqslant 30$	$\leqslant 1.5$			
Silt	$w>30$	Not considered			
Clay	$w_p+9<w \leqslant w_p+15$	$\leqslant 2.0$			
Clay	$w \geqslant w_p+15$	Not considered			

Notes: 1 w_p is the plastic limit and w is the average value of natural moisture content in frozen layer before freezing (%);
2 The saline permafrost is not included in the table;
3 When the plasticity index is greater than 22, the frost-heaving property shall be lowered by one level;
4 When the content of grains with a size less than 0.005 mm is greater than 60%, the soil is classified as non-frost-heaving soil;
5 The frost-heaving property of the gravel shall be determined according to the type of the fill when the fill content is greater than 40% of the total mass.
6 Aquiclude is in the active layer of seasonal freezing and seasonal thawing;
7 For works sensitive to frost-heaving deformation, the influence of the slight frost-heaving property of soil (classified as "no frost-heaving") on the works shall be analyzed.

4.3.8 The judgment and classification of fill shall comply with the following requirements:

1 The soil filled by human activities shall be judged as fill, which is generally characterized by complicated compositions and short consolidation time.

2 The classification of fill by the composition and filling method shall comply with Table 4.3.8.

Table 4.3.8 Classification of Fill

Name	Characteristics
Waste fill	Containing much construction waste, industrial waste, domestic waste and other impurities
Plain fill	Composed of gravel, sands, silt and clay, no impurities or few impurities
Dredger fill	Formed by hydraulic filling of mud or sand
Compacted fill	Tamped and compacted manually according to certain standards

Words Used for Different Degrees of Strictness

In order to mark the differences in executing the requirements in this Standard, words used for different degrees of strictness are explained as follows:

(1) Words denoting a very strict or mandatory requirement:

"Must" is used for affirmation; "must not" is used for negation.

(2) Words denoting a strict requirement under normal conditions:

"Shall" is used for affirmation; "shall not" is used for negation.

(3) Words denoting a permission of slight choice or an indication of the most suitable choice when conditions permit:

"Should" is used for affirmation; "should not" is used for negation.

(4) "May" is used to express the option available, sometimes with the conditional permit.